WILD BIOMES

ANIMALS OF THE AMAZON RAIN FOREST

by Mari Schuh

PEBBLE
a capstone imprint

Pebble Emerge is published by Pebble, an imprint of Capstone.
1710 Roe Crest Drive
North Mankato, Minnesota 56003
www.capstonepub.com

Library of Congress Cataloging-in-Publication Data
Title: Animals of the Amazon rain forest / by Mari Schuh.
Description: North Mankato, Minnesota : Pebble, an imprint of Capstone Press, [2022] | Series: Wild biomes | Includes bibliographical references and index. | Audience: Ages 6-8 | Audience: Grades 2-3 | Summary: "Biomes are home to unique animals and plants. Give beginning readers an introductory look at the Amazon Rain Forest! Readers will get an up-close look at the characteristics of the land and weather and how monkeys, parrots, anacondas, sloths, and other animals have adapted to life in this amazing biome"-- Provided by publisher.
Identifiers: LCCN 2020027289 (print) | LCCN 2020027290 (ebook) | ISBN 9781977131928 (library binding) | ISBN 9781977132949 (paperback) | ISBN 9781977153814 (pdf)
Subjects: LCSH: Rain forest animals--Amazon River Region--Juvenile literature.
Classification: LCC QL235 .S38 2022 (print) | LCC QL235 (ebook) | DDC 591.7340981/1--dc23
LC record available at https://lccn.loc.gov/2020027289
LC ebook record available at https://lccn.loc.gov/2020027290

Image Credits
Alamy: Nature Picture Library, 8, 14; Capstone: Eric Gohl, 5; Shutterstock: DeLoyd Huenink, 10, Dirk Ercken, 11, Erin Donalson, 13, Guillaume David, 12, Jelmer van Nuss, 17, Leonardo Mercon, 7, MarcusVDT, 9, Ondrej Prosicky, Cover, OSTILL is Franck Camhi, 15, reisegraf.ch, 6, Savejungle, 1, Svitlana-ua, 21, Tarcisio Schnaider, 19, worldclassphoto, 4

Design Elements
Capstone; Shutterstock: Savejungle

Editorial Credits
Editor: Jaclyn Jaycox; Designer: Hilary Wacholz; Media Researcher: Jo Miller; Production Specialist: Spencer Rosio

All internet sites appearing in back matter were available and accurate when this book was sent to press.

Printed in the United States 6148

Table of Contents

The Amazon Rain Forest 4

Rain Forest Animals 6

Life in the Rain Forest 16

Handprint Animal Art 20

Glossary 22

Read More 23

Internet Sites 23

Index 24

Words in **bold** are in the glossary.

THE AMAZON RAIN FOREST

The Amazon **rain forest** is full of life! Millions of different kinds of animals live there. Monkeys and birds sit high on tree branches. Big cats walk the forest floor.

This rain forest is the biggest in the world. It is in South America. The rain forest is warm. It is rainy and wet.

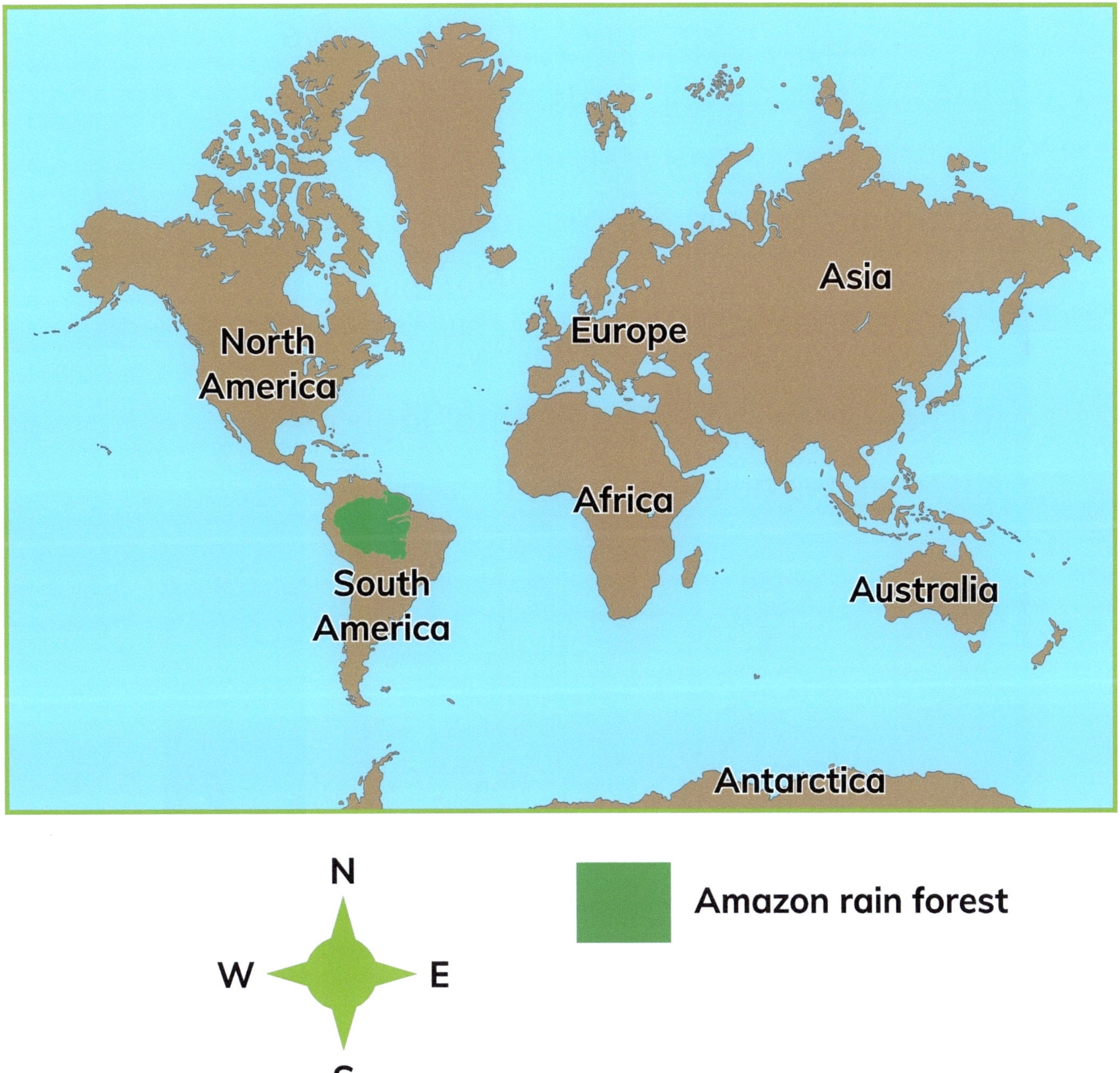
North America
Europe
Asia
Africa
South America
Australia
Antarctica
N
W
E
S
Amazon rain forest

RAIN FOREST ANIMALS

Jaguars hunt in the rain forest at night. They hunt alone. They eat deer, birds, and other animals. They hunt in water for fish, turtles, and crocodiles. Chomp! Jaguars' strong jaws help them eat their **prey**.

Jaguars' bodies help them in the rain forest. Spots on their fur help them hide. Big paws help them climb trees. Jaguars are great swimmers too.

Look at the harpy eagle's short wings. They help the eagle fly easily between the forest's trees. It also has huge **talons**. The eagle grabs sloths and monkeys. It grabs snakes and lizards too.

The harpy eagle is patient. It waits for prey. It can turn its head upside down. The eagle can see prey better!

Many kinds of birds are found in the rain forest. Macaws and other parrots live there. Toucans do too. Their beaks are strong. They crack open nuts and fruits.

Poison dart frogs are bright and colorful. The colors are a warning to **predators**. Stay away! These tiny frogs have **poison**. They are not safe to eat.

Monkeys sit high in trees. They hang. They climb and swing. Their strong arms and legs grab branches. Their tails do too! Monkeys eat leaves, fruits, and nuts.

See the sloth? It is slow. It has curved claws that grab branches. **Algae** live on its fur. The algae make the sloth look green. Sloths blend in with the leaves. Predators can't easily find them.

Anacondas are the biggest snakes in the world. They live in swamps and rivers. They hide. Their eyes peek out. They look for fish and turtles to eat. Their jaws open wide. They swallow prey whole!

Giant river otters live in the water too. Webbed feet help them swim. Their noses and ears close as they swim. Thick fur keeps them warm and dry.

LIFE IN THE RAIN FOREST

Rain forest animals need one another. They are part of a **food web**. Monkeys need fruits and leaves to eat. Harpy eagles need monkeys and other animals to eat. When too many animals die, other animals do not have food. Each plant and animal is important.

Rain forest animals face some dangers. People hunt animals there. Other people cut down trees. They clear the land. Roads and homes are built. Animals have fewer places to live.

The Amazon rain forest is important. People are working together. They are protecting the plants and animals that live there.

Handprint Animal Art

So many animals live in the rain forest! Try this fun activity to make a cheetah out of your handprint!

What You'll Need:

- paper
- paint
- paintbrush

What You'll Do:

1. Use a paintbrush to paint your hand. Make a handprint on a piece of paper. Turn the paper upside down so the fingers are at the bottom.
2. Draw or paint the cheetah's face on the thumb of your handprint. Draw the feet on the ends of the handprint's fingers.

3. Add a tail on the opposite side of your handprint's thumb.
4. Add spots all over the cheetah in a different color.
5. Add your own touch to it! You could show the cheetah hiding in tall grass. Be creative!

Glossary

algae (AL-jee)—small plants without roots or stems that grow in water or on damp surfaces

food web (FOOD WEB)—many food chains connected to one another

poison (POI-zuhn)—a substance that can kill or harm an animal or person

predator (PRED-uh-tur)—an animal that hunts other animals for food

prey (PRAY)—an animal that is hunted by other animals for food

rain forest (RAYN FOR-ist)—a thick forest where rain falls almost every day

talon (TAL-uhn)—a long, sharp claw

Read More

Delano, Marfé Ferguson. *Rain Forests.* Washington, D.C.: National Geographic Kids, 2017.

Gardeski, Christina Mia. *All About Rain Forests.* North Mankato, MN: Capstone Press, 2018.

Klepeis, Alicia. *Living in the Rain Forest.* New York: PowerKids Press, 2021.

Internet Sites

DK Findout!: Amazon Rain Forest
www.dkfindout.com/us/animals-and-nature/habitats-and-ecosystems/amazon-rain-forest/

National Geographic Kids: 10 Amazing Amazon Facts!
www.natgeokids.com/uk/discover/geography/physical-geography/amazon-facts/

Rainforest Alliance: Kids' Games and Activities
www.rainforest-alliance.org/kids

Index

algae, 13
anacondas, 14

birds, 4, 6, 10
branches, 4, 12, 13

clearing land, 18

fish, 6, 14
food webs, 16
fruits, 10, 12, 16

harpy eagles, 8–9, 16
homes, 18

jaguars, 6–7

leaves, 12, 13, 16

macaws, 10
monkeys, 4, 8, 12, 16

nuts, 10, 12

parrots, 10
poison dart frogs, 11

river otters, 15
roads, 18

sloths, 8, 13

talons, 8
toucans, 10
turtles, 6, 14